STRATEGIC TERRORISM

A CALL TO ACTION

NATHAN MYHRVOLD

The Lawfare Research Paper Series

RESEARCH PAPER NO. 2 – 2013

July 2013

Working Draft - Subject to Revision

ISBN 978-0-9827610-4-5

Cover image credit: Kasia Biel, photos.com

Contents

Introduction 4

The Power of the Stateless 6

Turning Life Against the Living 17

Tactical vs. Strategic Counterterrorism 28

The Audacity of Courage 40

References 53

Introduction

TECHNOLOGY CONTAINS NO INHERENT MORAL DIRECTIVE—it empowers people, whatever their intent, good or evil. This fact, of course, has always been true: when bronze implements supplanted those made of stone, the ancient world got swords and battle-axes as well as scythes and awls. Every technology has violent applications because that is one of the first things we humans ask of our tools.

The novelty of our present situation is that modern technology can provide small groups of people with much greater lethality than ever before. We now have to worry that private parties might gain access to weapons that are as destructive as—or possibly even more destructive than—those held by any nation-state. A handful of people, perhaps even a single individual, now have the ability to kill millions or even billions. Indeed, it is perfectly feasible, from a technological standpoint, to kill every man, woman, and child on Earth. The gravity of the situation is so extreme that getting the concept across without seeming silly or alarmist is challenging. Just thinking about the subject with any degree of seriousness numbs the mind.

Worries about the future of the human race are hardly novel. Indeed, the notion that terrorists or others might use weapons of mass destruction is so commonplace as to be almost passé. Spy novels, movies, and television dramas explore this plot frequently. We have become desensitized to this entire genre, in part because James Bond always manages to save the world in the end.

Reality may be different. In my estimation, the U.S. government, although well-meaning, is unable to protect us from the greatest threats we face. The other nations of the world are also utterly unprepared. Even obvious and simple steps are not being taken. The gap between what is necessary and what is being contemplated, much less being done, is staggering.

My appraisal of the present situation does not discount the enormous efforts of many brave men and women in law enforcement, intelligence services, and the military. These people are doing what they can, but the resources that we commit to defense and the gathering of intelligence are mostly squandered on problems that are far less dangerous to the American public than the ones we are ignoring.

Addressing the issue in a meaningful way will ultimately require large structural changes in many parts of the government. So far, however, our political leaders have had neither the vision to see the enormity of the problem nor the will to combat it. These weaknesses are not surprising: bureaucracies change only under extreme duress. And despite what some may say, the shocking attacks of September 11th, 2001, have not served as a wake-up call to get serious. Given the meager response to that assault, every reason exists to believe that sometime in the next few decades America will be attacked on a scale that will make 9/11 look trivial by comparison.

The goal of this essay is to present the case for making the needed changes *before* such a catastrophe occurs. The issues described here are too important to ignore.

The Power of the Stateless

For generations, the biggest menaces to our nation have been other nuclear-weapons states, especially the Soviet Union and China. Although the future of Russia seems to be of little concern to Americans these days, China, we are all told, will soon rival the United States as a great superpower. It will outgrow us. It will outspend us. It will dominate us.

That appraisal is certainly correct with respect to economic clout. But does China really pose a military threat? After all, it could not do something that would kill, say, a million Americans without expecting to lose 100 million Chinese in response. What's more, most of those million American victims would be wearing clothes, digital watches, and other consumer items made in China. Killing your best customers just isn't good business. The possibility of experiencing significant aggression from China thus seems very remote.

Nevertheless, *some* attention needs to be paid to the great People's Republic. Emotions over Taiwan's sovereign status run high enough that a future Chinese government might react poorly to a crisis there and could perhaps let events slip out of control. The traditional deterrence that keeps superpower ambitions in balance is thus important to maintain stability. But I suspect that China is a lot less dangerous than one might judge from the notice it currently receives.

Small, failed, bankrupt nation-states such as North Korea now pose the greater threat for both technological and political reasons. For one, countries with the least to lose have the greatest potential to do something nasty, particularly if, like North Korea, they hold nuclear weapons. Even if it never acts aggressively against the United States, each new entrant to the nuclear club becomes a potential point of dispersal of nuclear know-how to others. Libya's nuclear program was based on technology and materials from Pakistan and North Korea. The collapse of the Soviet Union has also greatly aided the dispersal of nuclear knowledge and potentially even complete weapons.

Nuclear bombs command enormous fear, which is why despots the world over see them as the ultimate and thus irresistible confirmation of status and power. Nuclear weapons in the hands of North Korea or Iran pose a threat to the United States, and it is tempting to think of our problems primarily in these terms—so tempting, indeed, that the United States is building an enormously expensive missile defense system to combat just such a peril.

Hostile nations with nuclear arms are clearly a source of concern. The problem they pose is an old one, however, and one that we are well equipped to manage.

The newer and less understood danger arises from the increasing likelihood that stateless groups bent on terrorism will gain access to nuclear weapons. Nuclear technology continues to spread to ever smaller and less organized countries. The Pakistani nuclear program, which fed the personal business venture of Abdul Qadeer Khan, a corrupt scientist who sold kits for each of the key areas of nuclear-weapons technology, demonstrates just how quickly control can be lost. The more countries that possess that technology, the more likely it is that lax oversight, irresponsible politics, or theft will put these terrible weapons into the hands of terrorists.

When this happens, the danger we now perceive to be coming from rogue states will pale in comparison. After all, because we usually don't know where they live, stateless groups wield much more *effective* destructive power because they can strike without fear of overwhelming retaliation.

Despite a $50-million reward for his capture, Osama bin Laden eluded us for more than a decade, and Ayman al-Zawahiri still remains at large, as does Mullah Mohammed Omar, the longtime leader of the Taliban. Although greatly weakened by the killing and capture of its commanders, the Taliban has fought the combined forces of the American and Afghan militaries to a stalemate in Afghanistan.

Other terrorist organizations have proved similarly resilient. The British military was never able to apprehend a sufficient number of Irish Republican Army leaders in Northern Ireland to quash their independence movement. Israel has similarly failed to assassinate enough of its key terrorist enemies to halt their activities, despite a very controversial and aggressive program that has killed large numbers of innocent civilians.

Even if one could figure out where terrorists were hiding, going after them and their supporters would probably not serve as a deterrent. Indeed, the fundamental equation of retaliation has become reversed because a frequent tactic of terrorists is to provoke reprisal attacks on their own people (or on the people they purport to represent) in the hope that the reprisals will sway popular opinion in their favor. How can the threat of a counter-attack be effective against enemies who want to provoke it? Among nation-states, the deterrent value of a retaliatory nuclear strike is enormous, but the United States is impotent when it comes to nuclear terrorism. Who would we target in response?

This question is hard enough to answer when seen through the lens of the 9/11 attacks, in which the enemy was recognizable, though hard to find. Right after the Twin Towers fell, angry Americans muttered that we should "just nuke the bastards." That kind of emotional reaction is more comprehensible when ethnicity and ideology make the division between "us" and "them" appear clearer than it is.

But think back a few years and consider the question of retaliation in the context of the second-worst terrorist attack in American history: the Oklahoma City bombing. Suppose Timothy McVeigh had committed a terrorist act against another country rather than against Oklahoma City. What would we have thought if a foreign military force had attacked McVeigh's hometown of Pendleton, New York? We would say, "Hey, wait a minute; that guy is a lone nutcase who does not represent anybody else in Pendleton." But that argument is just as true in Kandahar, Afghanistan or Falluja, Iraq. As tempting as it is to threaten tough reprisals, it demonstrably didn't work in those countries. You cannot stop terrorism that way.

The difficulty of reprisal gives terrorist groups enormous power. Terrorists can directly attack the most powerful country on Earth with an anonymity—and thus impunity—that no nation-state could match.

At most, a terrorist attack will provoke a military adventure in hostile territory that will prove much more costly to the superpower than to the terrorists. Indeed, provoking such attacks may be part of their plan.

The aftermath of 9/11 is a case in point. It seems likely that bin Laden and his cohorts did not anticipate that the United States would go so far as to invade Afghanistan. But the surprise was short-lived. After the United States invaded Afghanistan, Al Qaeda lost its host state and some personnel. Although that damage did slow it down a bit, it did not destroy the organization. Instead, the stateless Al Qaeda survived and adapted to its new status, even as the Taliban regime was forced into exile. The United States can claim some success against Al Qaeda in the years since 9/11, but it has hardly delivered its enemy a deathblow. The bottom line is clear: the United States has virtually no ability to deter attacks by stateless groups such as Al Qaeda.

Note that the key word here is "deter." We can retaliate to some degree with commando raids, by conducting drone strikes or even full scale invasions, as occurred in Afghanistan. These actions have had significant impact, and have helped reduce subsequent attacks. But the threat of counterattack does not act as an effective deterrence against terrorists. If we had been able to capture or kill the entire leadership, and completely disrupt the organization in short order following the 9/11 attacks, that might have deterred others. But as it stands, a would-be terrorist would conclude that it is extremely difficult to act against a stateless group, and thus would not be deterred.

Unfortunately, the fact is that Al Qaeda and Taliban have survived as organizations (along with many of their leaders) for more than a decade after 9/11. That does not make a strong case for deterrence. Indeed, future stateless groups could draw the opposite conclusion.

This difference in risks and returns between a superpower and a small adversary is sometimes called asymmetric warfare, but that simple phrase hardly captures the immense difficulties faced by the superior power. Those difficulties have grown in recent decades, and that trend seems likely to continue.

The most familiar form of asymmetric warfare is between conventional armies and guerilla fighters, who make adroit use of hit-and-run tactics. For more than half a century, guerilla warfare almost invariably meant barefoot peasants armed with AK-47s. Men equipped with little but courage and small arms could inflict tactical-level casualties and harass much larger armies. They could even, on occasion, embarrass superpowers into quitting an ill-planned conflict. Such forces never aspired to wield strategic levels of lethality, however. The Vietcong never hoped to topple Manhattan skyscrapers, nor did the Afghan mujahideen expect to level Moscow.

The Cold War was "cold" precisely because the Soviet Union could not risk directly waging an old-fashioned war of territorial expansion, the way Napoléon Bonaparte or Alexander the Great might have done. The U.S. reprisal would have been too much for them. Instead, the Soviets aided local Marxist guerilla groups, while we helped the other side. Superpower conflict by guerilla-warfare proxies was widespread from the 1950s through the demise of the Soviet Union.

This trend led to a generation of world leaders who came to power by commanding a guerilla movement; some even became heads of state. The path to power for a sufficiently ruthless and ambitious young man in most parts of the world was clear: lead a guerilla insurgency to "liberate" your country, forge a backroom alliance with one of the major powers, and become the local strongman.

Proxy fighting has taken a new form in the 21st century. Instead of proxy fighting by barefoot guerrillas against remote "U.S. interests," direct frontal assault on the citizens and military forces of the United States and other powerful countries is possible. Indeed, multiple examples of the success of this strategy exist. If a nation-state really wants to hurt the United States, why risk reprisal? Why not inflict damage by giving encouragement, resources, and direction to a group such as Al Qaeda?

Eventually, the world will figure out that stateless groups are more powerful than nation-states because terrorists can wield weapons and mount assaults that no nation-state would dare to attempt. Although this conclusion is only dimly perceived at present, the rising arc of terrorists' strategic advantage is clear and inexorable. Bin Laden, Abu Musab al-Zarqawi, and the Chechen terrorists, as well as their surviving followers, represent the vanguard. So far, they have limited themselves to dramatic and gory tactical terrorism: events such as 9/11, the butchering of Russian schoolchildren, decapitations broadcast over the Internet, and bombings in Madrid, London, and Boston. Strategic objectives cannot be far behind. Once stateless groups and their sponsors realize that they have unmatched power to threaten the mighty, incidents of terrorism will multiply in the same way that the guerrilla movements did during the Cold War.

The Technological Rise of Terrorism

THE QUICKEST PATH TO POWER FOR A RUTHLESS AND AMBITIOUS 21st-century man in many parts of the world is now to lead a stateless terror group. As a result, such groups will set the trend for international conflict throughout the next century. Sometimes they will be proxies for other combatants and will be covertly sponsored by nation-states. More often, they will follow their own agendas and be more lethal about it than any nation-state can afford to be.

Immunity from reprisal is only one of the trends shaping the rise of stateless power. Trends in technology are also relevant. Computers, the Internet, cellular and satellite telephones, and satellite TV give people unprecedented access to one another. This connectivity is mostly good. It enriches daily life and millions of lawful pursuits, including many that save lives. It also, however, enables a small group of dangerous people scattered around the world to organize themselves more effectively than ever before. Terrorist groups can now assemble a command-and-control structure that previously would have been available only to a wealthy nation-state.

Consider one example of how the proliferation of technology has tipped the playing field: U.S. spy satellites have in many crucial cases been rendered useless because our adversaries know when these eyes in the sky will pass overhead. Internet sites report amateur observations of their orbital parameters, computers enable the calculation of satellite positions, and cell phones are used to coordinate evasive actions. What was previously an unassailable technological advantage for the U.S. government has thus been greatly diluted by a combination of new consumer technologies. No amount of technology from the National Security Agency or National Reconnaissance Office allowed us to know with certainty whether bin Laden was in that compound in Abbottabad, Pakistan before the order to attack was given. Al-Zawahiri, Mullah Mohammed Omar, and other senior commanders of Al Qaeda and the Taliban have similarly managed to evade the most sophisticated and expensive manhunts in history. Determined men with disposable cell phones, laptops, and access to Third World Internet cafés can defeat our nation's multi-hundred-billion-dollar investment in space-based surveillance.

Communications technologies can also mobilize popular, political, and financial support for terrorists. Osama bin Laden's broadcasts on Al Jazeera, and Anwar al-Aulaqi's internet sermons were heard by millions of potential followers. The Boston Marathon bombers were inspired and, to a certain degree, trained by jihadist websites. How much more difficult would it be to recruit, organize, and fund Al Qaeda without that kind of reach? The bully pulpit afforded by modern communications has allowed what once would have been isolated fringe groups to knit together into formidable adversaries against the most powerful nations on Earth.

Indeed, such emerging loose coalitions may, in some ways, be more effective than large nation-states. Whenever the rate of technological change is high, new entrants have the upper hand. This phenomenon is well known in the commercial world. Young start-ups routinely challenge and displace companies that have been around for decades. The terrorists are playing the role of the dynamic start-up, and established nations such as the United States fillthe role of the old, slow, stuck-in-the-mud incumbents, unable to take advantage of the latest technological developments.

It is telling that on the tragic day of 9/11, the only effective defense came through adroit use of modern communications technology. Trillion-dollar Air Force defense systems sat largely idle and useless to defend the nation's capital. Meanwhile, passengers with cell phones and courage organized an unarmed assault on the terrorists in the cockpit and thus prevented United Airlines Flight 93 from taking more lives on the ground. This heroic story exemplifies the innovative use of new technology when the old, established approach fails. Unfortunately, technological innovation also works for the bad guys.

Communication has value to terrorists beyond command, control, and coordination. Terrorism works, after all, by instilling terror in large numbers of people. Although some terrorists do target victims directly, they also use them indirectly to frighten and manipulate a larger audience, and that aim requires mass communication. Terrorist groups seek the sensational because that is what advances their cause. The international media are, albeit reluctantly, the message-marketing-and-distribution department for today's terrorists. Tragedy anywhere reaches our living rooms with amazing speed and clarity.

The greatly increased reach that the media gives to contemporary terrorists strongly influences their choice of tactics. Successive attacks have to one-up the previous round. Competition to get the biggest splash on CNN and Al Jazeera will ultimately lead to an escalation and elaboration of terrorist acts. The spate of decapitations released on video on the Internet since 2002 raises a sad question: would the beheadings have occurred without a means to get the videos to millions of people?

The Internet and other communications technologies abet terrorist recruiting and fundraising as well. Countries finance their military ambitions through the power of the state to levy taxes and conscript young men. Stateless groups can't do the same, at least not in a conventional way, and that might appear to be a major limitation. But it's not.

Instead of levying taxes, non-state actors solicit donations on myriad Web pages and transfer funds via cell phone by exploiting informal *hawala* banking systems. Instead of drafting conscripts, they use the Internet to recruit the disaffected. CNN, Al Jazeera, YouTube, and many smaller websites carry the video advertising that hooks people into the funding and manpower networks.

Terrorism has always fed on publicity, but in decades past, its reach was very limited. In an information-poor world, the scare created by an act of terror was intensely local. To frighten a village, you would have to attack the village or, at least, one of its neighbors. The small geographic scope meant that terrorism worked only in the service of local conflicts. The short reach largely constrained it to being a purely tactical ploy.

The new range of contact afforded by modern communications changes the nature of terrorism in a qualitative way. Instead of being local, it is now global in its effects. From their tactical origins, terrorist attacks have been elevated to major strategic tools in the geopolitical game.

Modern communications technology even offers terrorist groups the hope of conquering lands in a 21st-century fashion and thus of achieving a goal that has frustrated conventional militaries. Saddam Hussein dreamed of conquest—and acted on the dream by waging war on both Iran and Kuwait. The results of these misadventures demonstrate the difficulties with old-fashioned wars of conquest. It is very easy to get mired in a stalemate, as happened to Hussein in Iran, or to be defeated simply by conventional military power, as happened to him in Kuwait. Indeed, in recent memory, no war of conquest anywhere in the world has been successful in capturing and permanently retaining territory.

But the aim of terrorists is somewhat different. They do not seek to capture territory per se but rather to capture power over a population through intimidation or mass murder. For example, both bin Laden and the Taliban are reputed to have sought restoration of an Islamic emirate in Afghanistan and, eventually, across the entire Middle East. This is an incredibly audacious objective. It would be crazy for any of the nation-states in the region to contemplate wars of conquest to restore the caliphate to its eighth-century glory. They would be rebuffed or repelled long before they could complete the task.

But modern communications gives Al Qaeda and the Taliban a chance to establish a virtual emirate at the very least. Mullah Omar and his organization already operate one of the most influential virtual governments in the region. To the extent that competition exists, it does not come from the reigning government of Afghanistan, which commands little popular support. Instead, the leaders with the greatest celebrity and influence in the region are people such as Hassan Nasrallah, leader of the stateless Hezbollah organization. The incredible reach of modern communications gives al-Zawahiri, Nasrallah, and their ilk an influence that crosses borders and transcends the local political structure. Although resurrecting a caliphate may elude stateless actors, the influence they command makes them more powerful in many ways than the leaders of nation-states.

This state of affairs is the new world order enabled by technology. Stateless groups can recruit and organize followers across national borders. They can also develop the lethality sufficient to scare anyone, superpowers included. For the first time, guerilla warfare can be fought at the strategic level. This capability gives terrorist groups the power to threaten the established community of nations, including the remaining superpower, in a truly unique way. The net result is that terrorists can deploy more *usable* military power than any nation-state. It will take some time for these groups to fully exploit this advantage, but the stage is set for the stateless to become the greatest threats and thus the pre-eminent military players on the globe.

Indeed, we seem to be entering the golden age of stateless organizations. During this age, the military supremacy and political influence of nation-states will be challenged by much smaller groups that can wield both political influence and power with cruelty and without the apparatus of a state. As a result, massive terrorist attacks like 9/11—as well as low-level events such as suicide bombings, kidnappings, and assassinations—will occur with greater frequency.

Bad as that is, it is unfortunately only part of the story. The organizational power of communications and computing has its destructive limits because such technology is not itself lethal. The crucial additional factor is that weapons technology is becoming more accessible and powerful every year.

The Democratization of Death Dealing

THROUGHOUT HISTORY, THE LETHALITY OF WEAPONS technology has inexorably increased. Bronze weapons were better than those made of stone; steel later outdid bronze; guns replaced bows; and so forth. Each new generation of weapons technology was more lethal than its predecessor.

Yet a general rule prevailed: successively more lethal weapons required successively larger investments and industrial bases. Making a bronze sword involved mining, smelting, and casting. Making a steel sword required forced-air furnaces to melt the iron, alloying technology to produce the steel, and forging techniques to shape the blade.

This trend continued unabated through modern history. The outcome of World War II was decided in large part by the superiority of U.S. industrial output to that of Germany and Japan. Heroism and courage are great things, but ammunition, tanks, and ships ultimately matter more.

Nuclear weapons were the zenith of this arc of increasing lethality and effort. A single device could destroy an entire city, but it also cost as much as an entire city and was far harder to build. The first nuclear explosives were created by the three-year Manhattan Project, which at its peak employed 130,000 people. It cost more than $2 billion in the currency of the time—the equivalent, in 2013, of more than $26 billion. But that is just money. To put the engineering and industrial effort in perspective, the project became comparable in manpower and capital cost to the entire prewar U.S. automobile industry. Yet these measures underestimate the true cost and difficulty of the Manhattan Project because the special circumstances of the war inspired the necessary talent to volunteer. Another factor is nuclear safety—in those days a combination of naïveté and bravery during a war meant that many people took incredible personal risks—and paid the price with their lives or health.

From that time forward, nuclear weapons put an enormous strain on the economies of the states that decided to field them; in addition to the bombs themselves, the expense of the delivery systems and other aspects of nuclear warfare required a staggering investment. The escalating cost of competing with U.S. weapons systems, particularly the "Star Wars" missile-defense system, is widely cited as one of the factors in the economic collapse of the Soviet Union.

The technological sophistication required to build and maintain these weapons is also daunting. This level of sophistication is the primary reason that only two countries became nuclear superpowers. A larger number of states produced some nuclear weapons but did not ante into the full superpower game.

The cost of nuclear weapons had two stabilizing effects. First, the set of entities that could wreak nuclear havoc was very small. Second, each leader with a finger on "the button" tended to bear the full responsibility for a large and complex state—each understood that use of the weapons would bring a very dangerous reprisal, which helps explain why the Cold War stayed cold. The inescapable equation tying highly lethal weapons systems to high cost and complexity meant that the power to devastate was available only to the richest and most sophisticated states—until now.

For the first time in human history, the curve of cost versus lethality has turned rapidly downward, falling many orders of magnitude in just a generation. Today, tremendously lethal technology is available on the cheap. Anyone—even a stateless group—can have the deadliest weapons on Earth. Several trends led to this inflection point. One is nuclear proliferation, which in recent years reached a tipping point at which access to nuclear weapons became impossible to control or limit in any absolute way. The collapse of the Soviet Union scattered ex-Soviet weapons across many poorly governed and policed states, and from there, the weapons may spread further into the hands of terrorists. At the same time, the set of ragtag countries that have developed homegrown nuclear devices is large and growing. The entrance to the nuclear-weapons club, once limited to a small number of sophisticated and stable countries, is now far more open.

It is only a matter of time before a nuclear bomb gets into the hands of a terrorist group, whether by theft or construction. A nuclear weapon smuggled into an American city could kill between 100,000 and 1,000,000 people, depending on the nature of the device, the location of ground zero, and the altitude of detonation. An optimist might say that it will take another decade for such a calamity to take place; a pessimist would point out that the plot may already be under way.

Chemical weapons, particularly nerve agents, are another new addition to the terrorist arsenal. Sarin, a frighteningly lethal poison discovered in 1938 and stockpiled (although never used) by the Nazis, was produced and released in locations in the Tokyo subway system in 1995 by Aum Shinrikyo, a Japanese religious cult. The attack injured nearly 3,800 people and killed 12. A botched distribution scheme in the Tokyo subway spared many of the intended victims; better dispersal technology would have resulted in a vastly higher death toll.

Cult members had more morbid ambitions than a subway attack. They had gathered hundreds of tons of raw materials and had procured a Russian military helicopter to use in spraying the nerve agent over Tokyo. Experts have estimated that Aum Shinrikyo had the ingredients to produce enough sarin to kill millions of people in an all-out attack. The civil war in Syria, whose military is known to possess stockpiles of sarin and other chemical weapons, raises the prospect that these munitions could fall into the hands of extremists.

Frightening as such possibilities are, nuclear bombs and chemical agents pale in lethality when compared with biological weapons. Indeed the term "weapon" is not entirely adequate because biological agents include not only pathogens that are controllable (in the traditional sense) but also those that are not.

Even more so than with nuclear weapons, the cost and technical difficulty of producing biological arms has dropped precipitously in recent decades with the boom in industrial molecular biology. A small team of people with the necessary technical training and some cheap equipment can create weapons far more terrible than any nuclear bomb. Indeed, even a single individual might do so.

Taken together, these trends utterly undermine the lethality-versus-cost curve that existed throughout all of human history. Access to extremely lethal agents—even to those that may exterminate the human race—will be available to nearly anybody. Access to mass death has been democratized; it has spread from a small elite of superpower leaders to nearly anybody with modest resources. Even the leader of a ragtag, stateless group hiding in a cave—or in a Pakistani suburb—can potentially have "the button."

Turning Life Against the Living

THE FIRST AND SIMPLEST KINDS OF BIOLOGICAL WEAPONS are those that are not contagious and thus do not lead to epidemics. These have been developed for use in military conflicts for most of the 20th century. Because the pathogens used are not contagious, they are considered controllable: that is, they have at least some of the command-and-control aspects of a conventional weapon. Typically, these pathogens have been "weaponized," meaning bred or refined for deployment by using artillery shells, aerial bombs, or missiles much like conventional explosive warheads. They can be highly deadly.

Anthrax is the most famous example. In several early-20th-century outbreaks, it killed nearly 90% of those infected by inhaling bacterial spores into their lungs. Anthrax was used in the series of mail attacks in the United States in the fall of 2001. Even with advanced antibiotic treatment, 40% of those who contracted inhalational anthrax died during the 2001 attacks.[1]

That crime is believed to have been the work of a lone bioweapons scientist who sought to publicize the threat of a biological attack and boost funding for his work on anthrax vaccines. This conclusion is consistent with the fact that virtually no effort was made to disperse the bacterium—indeed, the letters carrying the spores thoughtfully included text warning of anthrax exposure and recommending that the recipient seek immediate treatment. Despite this intentional effort to limit rather than spread the infection, a surprising amount of trouble was caused when the fine anthrax powder leaked from envelopes and contaminated other mail. Before this episode, nobody would have guessed that letters mailed in New Jersey to addresses in Manhattan and Washington, D.C., could kill someone in Connecticut, but they did. And no one would have predicted that a domestic bioterrorist launching multiple attacks, including one against the U.S. Congress, would elude the FBI for years. But that is what happened.

What if such an attack were made not by some vigilante trying to alert the world to the dangers of bioweapons but instead by a real sociopath? Theodore J. Kaczynski, better known as the "Unabomber," may have been such a person. He was brilliant enough to earn a Ph.D. in mathematics from the University of Michigan yet was mentally disturbed enough to be a one-man terrorist cell: His mail bombs claimed victims over nearly two decades. Kaczynski certainly had enough brains to use sophisticated methods, but because he opposed advanced technology, he made untraceable low-tech bombs that killed only three people. A future Kaczynski with training in microbiology and genetics, and an eagerness to use the destructive power of that science, could be a threat to the entire human race.

Indeed, the world has already experienced some true acts of biological terror. Aum Shinrikyo produced botulinum toxin and anthrax and reportedly released them in Tokyo on four separate occasions. A variety of technical and organizational difficulties frustrated these attacks, which did not cause any casualties and went unrecognized at the time for what they were, until the later Sarin attack clued in the authorities.[2] Had the group been a bit more competent, things could have turned out far worse.

One 2003 study found that an airborne release of one kilogram of an anthrax-spore-containing aerosol in a city the size of New York would result in 1.5 million infections and 123,000 to 660,000 fatalities, depending on the effectiveness of the public health response.[3] A 1993 U.S. government analysis determined that 100 kilograms of weaponized anthrax, if sprayed from an airplane upwind of Washington, D.C., would kill between 130,000 and three million people.[4] Because anthrax spores remain viable in the environment for more than 30 years,[1] portions of a city blanketed by an anthrax cloud might have to be abandoned for years while extensive cleaning was done. Producing enough anthrax to kill 100,000 Americans is far easier to do—and far harder to detect—than is constructing a nuclear bomb of comparable lethality.

Anthrax, moreover, is rather benign as biological weapons go. The pathogen is reasonably well understood, having been studied in one form or another in biowarfare circles for more than 50 years. Natural strains of the bacterium are partially treatable with long courses of common antibiotics such as ciprofloxacin if the medication is taken sufficiently quickly, and vaccination soon after exposure seems to reduce mortality further.[5]

But bioengineered anthrax that is resistant to both antibiotics and vaccines is known to have been produced in both Soviet and American bioweapons laboratories. In 1997, a group of Russian scientists even openly published the recipe for one of these superlethal strains in a scientific journal.[6]

In addition, numerous other agents are similar to anthrax in that they are highly lethal but not contagious. The lack of contagion means that an attacker must administer the pathogen to the people he wishes to infect. In a military context, this quality is generally seen as a good thing because the resulting disease can be contained in a specific area. Thus, the weapon can be directed at a well-defined target, and with luck, little collateral damage will result.

Unfortunately, many biological agents are communicable and so can spread beyond the people initially infected to affect the entire population. Infectious pathogens are inherently hard to control because there is usually no reliable way to stop an epidemic once it starts. This property makes such biological agents difficult to use as conventional weapons. A nation that starts an epidemic may see it spread to the wrong country—or even to its own people. Indeed, one cannot target a small, well-defined population with a contagious pathogen; by its nature, such a pathogen may infect the entire human race.

Despite this rather severe drawback, both the Soviet Union and the United States, as well as Imperial Japan, investigated and produced contagious bioweapons. The logic was that their use in a military conflict would be limited to last-ditch, "scorched earth" campaigns, perhaps with a vaccine available only to one side.

Smallpox is the most famous example. It is highly contagious and spreads through casual contact. Smallpox was eradicated in the wild in 1977, but it still exists in both U.S. and Russian laboratories, according to official statements.[7] Unofficial holdings are harder to track, but a number of countries, including North Korea, are believed to possess covert smallpox cultures.

Biological weapons were strictly regulated by international treaty in 1972. The United States and the Soviet Union agreed not to develop such weapons and to destroy existing stocks. The United States stopped its bioweapons work, but the Russians cheated and kept a huge program going into the 1990s, thereby producing thousands of tons of weaponized anthrax, smallpox, and far more exotic biological weapons based on genetically engineered viruses. No one can be certain how far either the germs or the knowledge has spread since the collapse of the Soviet Union.

Experts estimate that a large-scale, coordinated smallpox attack on the United States might kill 55,000 to 110,000 people, assuming that sufficient vaccine is available to contain the epidemic and that the vaccine works.[8, 9] The death toll may be far higher if the smallpox strain has been engineered to be vaccine-resistant or to have enhanced virulence.

Moreover, a smallpox attack on the United States could easily broaden into a global pandemic, despite the U.S. stockpile of at least 300 million doses of vaccine. All it would take is for one infected person to leave the country and travel elsewhere. If New York City were attacked with smallpox, infections would most likely appear on every continent, except perhaps Antarctica, within two weeks. Once these beachheads were established, the epidemic would spread almost without check because the vaccine in world stockpiles and the infrastructure to distribute it would be insufficient. That is particularly true in the developing world, which is ill equipped to handle their current disease burden to say nothing of a return of smallpox. Even if "only" 50,000 people were killed in the United States, a million or more would probably die worldwide before the disease could be contained, and containment would probably require many years of effort.

As horrible as this would be, such a pandemic is by no means the worst attack one can imagine, for several reasons. First, most of the classic bioweapons are based on 1960s and 1970s technology because the 1972 treaty halted bioweapons development efforts in the United

States and most other Western countries. Second, the Russians, although solidly committed to biological weapons long after the treaty deadline, were never on the cutting edge of biological research. Third and most important, the science and technology of molecular biology have made enormous advances, utterly transforming the field in the last few decades. High school biology students routinely perform molecular-biology manipulations that would have been impossible even for the best superpower-funded program back in the heyday of biological-weapons research. The biowarfare methods of the 1960s and 1970s are now as antiquated as the lumbering mainframe computers of that era. Tomorrow's terrorists will have vastly more deadly bugs to choose from.

Consider this sobering development: in 2001, Australian researchers working on mousepox, a nonlethal virus that infects mice (as chickenpox does in humans), accidentally discovered that a simple genetic modification transformed the virus.[10, 11] Instead of producing mild symptoms, the new virus killed 60% of even those mice already immune to the naturally occurring strains of mousepox. The new virus, moreover, was unaffected by any existing vaccine or antiviral drug. A team of researchers at Saint Louis University led by Mark Buller picked up on that work and, by late 2003, found a way to improve on it: Buller's variation on mousepox was 100% lethal, although his team of investigators also devised combination vaccine and antiviral therapies that were partially effective in protecting animals from the engineered strain.[12, 13] Another saving grace is that the genetically altered virus is no longer contagious. Of course, it is quite possible that future tinkering with the virus will change that property, too.

Strong reasons exist to believe that the genetic modifications Buller made to mousepox would work for other poxviruses and possibly for other classes of viruses as well. Might the same techniques allow chickenpox or another poxvirus that infects humans to be turned into a 100% lethal bioweapon, perhaps one that is resistant to any known antiviral therapy? I've asked this question of experts many times, and no one has yet replied that such a manipulation couldn't be done.

This case is just one example. Many more are pouring out of scientific journals and conferences every year. Just last year, the journal

Nature published a controversial study done at the University of Wisconsin–Madison in which virologists enumerated the changes one would need to make to a highly lethal strain of bird flu to make it easily transmitted from one mammal to another.[14]

Biotechnology is advancing so rapidly that it is hard to keep track of all the new potential threats. Nor is it clear that anyone is even trying. In addition to lethality and drug resistance, many other parameters can be played with, given that the infectious power of an epidemic depends on many properties, including the length of the latency period during which a person is contagious but asymptomatic. Delaying the onset of serious symptoms allows each new case to spread to more people and thus makes the virus harder to stop.

This dynamic is perhaps best illustrated by HIV, which is very difficult to transmit compared with smallpox and many other viruses. Intimate contact is needed, and even then, the infection rate is low. The balancing factor is that HIV can take years to progress to AIDS, which can then take many more years to kill the victim. What makes HIV so dangerous is that infected people have lots of opportunities to infect others. This property has allowed HIV to claim more than 30 million lives so far, and approximately 34 million people are now living with this virus and facing a highly uncertain future.[15]

A virus genetically engineered to infect its host quickly, to generate symptoms slowly—say, only after weeks or months—and to spread easily through the air or by casual contact would be vastly more devastating than HIV. It could silently penetrate the population to unleash its deadly effects suddenly. This type of epidemic would be almost impossible to combat because most of the infections would occur before the epidemic became obvious.

A technologically sophisticated terrorist group could develop such a virus and kill a large part of humanity with it. Indeed, terrorists may not have to develop it themselves: some scientist may do so first and publish the details.

Given the rate at which biologists are making discoveries about viruses and the immune system, at some point in the near future, someone may create artificial pathogens that could drive the human race to extinction. Indeed, a detailed species-elimination plan of this nature was openly proposed in a scientific journal.

The ostensible purpose of that particular research was to suggest a way to extirpate the malaria mosquito, but similar techniques could be directed toward humans.[16] When I've talked to molecular biologists about this method, they are quick to point out that it is slow and easily detectable and could be fought with biotech remedies. If you challenge them to come up with improvements to the suggested attack plan, however, they have plenty of ideas.

Modern biotechnology will soon be capable, if it is not already, of bringing about the demise of the human race—or at least of killing a sufficient number of people to end high-tech civilization and set humanity back 1,000 years or more. That terrorist groups could achieve this level of technological sophistication may seem far-fetched, but keep in mind that it takes only a handful of individuals to accomplish these tasks. Never has lethal power of this potency been accessible to so few, so easily. Even more dramatically than nuclear proliferation, modern biological science has frighteningly undermined the correlation between the lethality of a weapon and its cost, a fundamentally stabilizing mechanism throughout history. Access to extremely lethal agents—lethal enough to exterminate *Homo sapiens*—will be available to anybody with a solid background in biology, terrorists included.

The 9/11 attacks involved at least four pilots, each of whom had sufficient education to enroll in flight schools and complete several years of training. Bin Laden had a degree in civil engineering. Mohammed Atta attended a German university, where he earned a master's degree in urban planning—not a field he likely chose for its relevance to terrorism. A future set of terrorists could just as easily be students of molecular biology who enter their studies innocently enough but later put their skills to homicidal use. Hundreds of universities in Europe and Asia have curricula sufficient to train people in the skills necessary to make a sophisticated biological weapon, and hundreds more in the United States accept students from all over the world.

Thus it seems likely that sometime in the near future a small band of terrorists, or even a single misanthropic individual, will overcome our best defenses and do something truly terrible, such as fashion a bioweapon that could kill millions or even billions of people. Indeed, the creation of such weapons within the next 20 years seems to be a virtual certainty. The repercussions of their use are hard to estimate. One approach is to look at how the scale of destruction they may cause compares with that of other calamities that the human race has faced.

The Grim Calculus of Mass Mortality

GRAPPLING WITH THE MIND-NUMBING STATISTICS OF MASS death is nearly impossible. Some 56 million people died in World War II, for example, whereas "only" 15 million died in World War I. The human misery associated with such a catastrophe cannot be captured in the figures alone, yet we cannot escape the logic that threats must be measured and prioritized.

Using a logarithmic scale to count fatalities by powers of 10 can encompass the large range of possibilities. Thus, an event that kills 1,000, or 10^3, people would be magnitude 3, or M3.0 for short. On this scale, World War II is an M7.7 event, and World War I is M7.2. Table 1 on the next page gives some examples.

The first thing that is apparent from these tabulated values is that some causes of death have much greater psychological impact than others, regardless of the number of people killed. For example, 9/11 ranks below annual U.S. traffic deaths—indeed, almost as many people perish in a typical *month* on American highways as died in the 9/11 attacks. So why were we so worked up? We have had more than 100 years to become used to the fact that roads are dangerous and we thus expect a certain level of risk when traveling on them. Therefore, when a single dramatic and totally unexpected event kills almost 3,000 innocent people at once, the impact on society is far greater than the slow losses from more mundane causes.

Plane crashes are an example of a cause of death that receives an enormous amount of public attention, even though commercial aircraft generate a factor of 3,000 fewer fatalities worldwide than do

EVENT OR POPULATION	POP. OR FATALITIES	MAGNITUDE
Total World Population	7,000,000,000	9.8
Population Of China	1,350,000,000	9.1
Population Of The United States	313,000,000	8.5
HIV/AIDS Cumulative Deaths + Currently Infected	64,000,000	7.8
World War II, Total	56,125,262	7.7
Influenza Pandemic Of 1918, Total	20,000,000	7.3
World War I, Total	14,958,886	7.2
Deaths In U.S. From All Causes In 2011	2,468,435	6.4
Vietnam Conflict, Total	1,900,000	6.3
Aids Deaths In 2011	1,700,000	6.2
Worldwide Annual Traffic Deaths In 2011	1,400,000	6.1
Rwandan Genocide Of 1994-1995	1,000,000	6.0
Influenza Epidemic Of 1918 (U.S. Only)	675,000	5.8
World War Ii (U.S. Only)	500,000	5.7
Indian Ocean Tsunami Of 2004	230,000	5.4
World War I (U.S. Only)	116,516	5.1
Nuclear Bombing Of Hiroshima	90,000	5.0
Vietnam Conflict (U.S. Only)	58,153	4.8
Traffic Deaths In 2011 (U.S. Only)	29,757	4.5
Murders In 2011 (U.S. Only)	14,612	4.2
September 11Th Terrorist Attack	2,996	3.5
Aircraft Crash Deaths In 2011 (U.S. Only)	494	2.7

TABLE 1. Relative magnitudes of human populations and mass fatalities, expressed as a power of 10. Magnitudes of U.S.-only events were calculated by using the U.S. population at the time.

Sources: U.S. Census Bureau, UNAIDS, Wikipedia, U.S. National Center for Health Statistics, U.S. Department of Transportation, Federal Bureau of Investigation, National Transportation Safety Board, U.S. Department of Defense automobiles and account for a negligible percentage of total deaths. Yet such events loom larger in our imaginations than these statistics reflect; airplane accidents are feared much more than they should be, rationally speaking. One reason is that an airplane accident is high drama. It is also unexpected and outside our control as passengers. We are prepared to accept our own culpability for overeating ourselves to death or for the highway deaths that result from our driving errors. In a plane crash, however, lots of people die in a spectacular and gory fashion without much sense of their own responsibility for the tragedy. Death coming as a bolt from the blue tends to get our attention. This attention is the same sort that terrorists wish to inspire through dramatic acts of violence. With this bias in mind, we can look more closely at the magnitude of casualties that may result from terrorist attacks and natural pandemics.

Small events at the M1.0 to M2.0 level—such as suicide bombings or shooting rampages in shopping malls—would be traumatic if they proliferated, particularly at first. But if repeated events of this nature occurred over a long period of time, Americans would probably become inured to them, much as they have become accustomed to traffic accidents. This change in mind-set happened in Northern Ireland and Israel, where low-level terrorism became a way of life during the conflicts there that lasted many decades—and the same thing has happened more recently in Iraq and Afghanistan.

This phenomenon of acclimatization is also evident in crime statistics. In 2011, 14,612 murders occurred in the United States, but as recently as 1991, 24,703 occurred. Although homicide is lamentable, Americans seem to be able to handle 10,000 or so more deaths per year than we have recently experienced. One reason is that the overall rate is low—fewer than five people per 100,000. The other reason is that homicides generally occur as isolated instances: a murder here, a murder there. If instead, 1,000 or 10,000 died in a single event, it would be much scarier.

Similarly, if annual traffic deaths increased by 10,000 (M4.0), I suspect that few would even notice. But if the nation lost 10,000 people at once in a terrorist act, the reaction would be enormous. And what would happen if an M5.0 (100,000 dead) terrorist event took place in New York

TERRORIST EVENT	LOW M	HIGH M
Extinction of *Homo sapiens*	9.8	10
End of civilization	9.5	10
Pandemic from engineered bioweapon	6.0	10
Smallpox attack	6.0	8.0
Natural virus pandemic	6.0	8.0
Natural virus pandemic (U.S. only)	5.0	7.0
Smallpox attack (U.S. only)	4.7	6.5
Anthrax attack	4.0	6.5
Neurotoxin attack	5.0	6.0
Nuclear weapon	5.0	6.0
Dirty bomb	3.0	4.0
Aircraft as weapon	3.0	4.0
Truck bombing	2.0	3.0
Suicide bombing	1.0	2.5
Shooting rampage	1.0	1.5

TABLE 2. Range of magnitudes of fatalities plausible for terrorist attacks of various kinds.

or another major city? This magnitude corresponds to almost twice the number of Americans killed in the Vietnam conflict. The disruption to the country would make 9/11 look mild in comparison.

To put it in perspective, a single M5.0 event would kill more people than were killed cumulatively in all terrorist actions by all parties throughout history. It might take anywhere from 1,000 to 10,000 typical suicide bombings to equal it.

Which would be easier to perpetrate? One M5.0 event, or thousands of individual attacks? All factors point to one large strike, which, depending on what was done, could very well push the death toll toward M6.0.

At some point, terrorists will figure this out. As perverse as it may seem, their current obsession with small-scale suicide bombings may be a blessing in disguise. These attacks keep their cause in the news at a relatively modest cost in lives lost. A wave of suicide bombings in American shopping malls would be a terrible thing—but a lot less terrible than an event in which terrorists used a weapon of mass destruction to kill a far larger number of people.

Would They Do It?

COULD THIS REALLY HAPPEN? WOULD TERRORISTS REALLY TRY to kill millions of Americans? Or will they stick to convincing their own youth to blow themselves up in small-scale suicide bombings? It is one thing to chart the trends in the use of lethal technology by stateless groups, but the key question is "Should we really take this threat seriously?"

One might argue that terrorism will remain small in scale and that high-M events will not occur because terrorists will remain satisfied with scattered suicide bombings and other low-M carnage. Several lines of reasoning suggest that this is not the case and that stateless terror groups will acquire and use weapons having high-M impact.

The first argument is that stateless groups now have the same level of ambition as nation-states and ought to be treated as operating on the same footing. Was it rational to worry that the Soviet Union would launch a nuclear war to further their communist hegemony or simply to destroy the United States—or out of fear that we would attack them in this way first? Dealing with those questions consumed $1 trillion dollars of defense spending and shaped the Cold War. Thankfully, the dreaded nuclear exchange never came to pass, but we certainly took the possibility seriously. Indeed, retired cold warriors will tell you that it didn't occur precisely because we took it seriously.

The risk that Al Qaeda or some future group will use equally terrible weapons seems higher on every level. Its geopolitical goals are, if anything, more ambitious than the Soviets' were. Al Qaeda's ideology is more extreme. The group's vulnerability to counterattack or reprisal is far lower than anything the Soviets faced—it has already survived the worst our nation can throw at it. The terrorists have demonstrated a shocking degree of ruthlessness. Under any theory of risk, these foes must be considered more likely to act than the Soviets ever were.

Another reason terrorists would attack is the oldest justification in the world—because we're trying to get them. It's no secret that the United States aims to exterminate Al Qaeda and similar terrorist groups—and rightly so. With revenge and self-preservation on their minds, our primary adversaries are not likely to show us unnecessary mercy.

A more mundane reason to worry is that the information cascade that empowers stateless groups will ultimately demand more numerous and spectacular demonstrations of power to feed popular interest. Terrorism survives by making a big impact, and when the world gets desensitized to beheadings, the temptation to one-up the last attack increases.

Similarly, the arc of terrorism in Iraq—which spiked dramatically from 2004 through 2007 and then leveled off, only to resurge somewhat recently—may foreshadow an increasing risk to the United States. Terrorists quite rationally sought to destabilize Iraq and Afghanistan as a way to humble the United States and influence its policy by forcing a pullout. That strategy focused terrorists' attention more on these countries and possibly distracted some groups from directly attacking U.S. territory.

As U.S. forces withdraw from the region, these targets become less interesting. What next? Al Qaeda and other stateless groups will seek to build on their previous successes. They have successfully carved out a safe haven for themselves in the lawless frontiers of Pakistan. Dramatic attacks on the American homeland would be a natural next step.

The decentralized nature of stateless organizations raises another set of concerns. Once mass death becomes accessible to small groups, it is unclear who would be in control. This lack of direction has already been seen in various Al Qaeda attacks in Saudi Arabia and Europe, some of which clearly hurt the cause of Islamic terrorists. They took place because no single chain of command exists in the overall movement—it is, at best, a loose confederacy.

An additional issue might be called the "craziness factor." Small groups can have crazy goals. The smaller the group, the crazier they may be. The apocalyptic death cult Aum Shinrikyo is a case in point. Kaczynski is another example.

The belief that terror groups will not use terrible weapons if they get them seems foolish in the extreme. To borrow a phrase from *A Streetcar Named Desire,* to hold this belief is, in effect, to rely "on the kindness of" terrorists. Any rational analysis must assign a substantial amount of the terror risk to large-scale, high-magnitude events. Yet that is not how our defenses are organized and not how we are spending our resources. Instead, we focus most of our counterterrorism efforts on thwarting small-scale attacks—by, for example, confiscating grandma's four-ounce bottle of hand lotion at the airport.

Tactical vs. Strategic Counterterrorism

THE ENORMOUS RANGE OF POSSIBLE TERRORIST ACTIONS mirrors a situation encountered in modern warfare: the distinction between strategic and tactical engagements. Military commanders must confront war at many levels, from hand-to-hand combat to global thermonuclear war. That broad range is very difficult to cover with a single organization. The military answer is to split the problem into pieces by both scale and approach. The division by scale is usually phrased as the difference between strategic and tactical, whereas the division by approach is normally based on ground, sea, and air.

For example, Strategic Air Command is responsible for bomber- and missile-based strategic nuclear war. Strategic Air Command is largely separate from the rest of the Air Force; indeed, it was set up a year before the Air Force was founded. The Emergency War Powers Act placed the Strategic Air Command under the direct operational control of the president. A similar division of responsibility separates submarines that carry nuclear weapons from the rest of the Navy.

Running a modern military organization that did not have this sort of division between strategic and tactical activities would be impossible. The range of technologies and difference in approach are simply too great to treat conventional and nuclear threats as a single problem—it would be foolhardy to do so. Yet when it comes to terrorism, we are doing exactly that.

Tactical terrorism is important to fight. We want to keep hijackers off airplanes and suicide bombers out of shopping malls. Referring to such problems as tactical does not suggest they are unimportant. Rather, it highlights the need to make even greater efforts to thwart strategic terrorism.

In the case of terrorist threats, the division into strategic and tactical could be made, for example, by designating attacks with a potential of M5.0 and above as strategic and the rest as tactical. One could choose a different threshold, but M5.0 covers nuclear weapons and serious biological attacks. A higher threshold would leave small nukes out of the picture. A lower threshold would widen the strategic focus with smaller attacks.

Alternatively, we could distinguish by the technology used. Nuclear, chemical, and biological threats (weapons of mass destruction) would be considered strategic; attacks with conventional weapons would be deemed tactical.

However it is done, separating strategic and tactical terrorist threats is crucial because we must use very different techniques, personnel, and methods to combat the two. Conventional security and law-enforcement activities—such as metal detectors at airports and air marshals onboard commercial planes—can largely handle the tactical threat.

Strategic counterterrorism is another matter altogether. The security forces inside the U.S. are, generally speaking, useless against terrorists intent on using contagious biological agents or nuclear weapons. By the time such terrorists have arrived at the airport or harbor, they have all but won. An airport is a great place to infect people without even boarding a plane, and a harbor can be an ideal location to set off a nuclear bomb hidden in a shipping container.

Are U.S. authorities doing enough to combat terrorism at the strategic level? The indirect evidence indicates that the answer is most certainly no. Aside from a few inadequate efforts to screen a fraction of ships and aircraft overseas before they depart for American shores, the problem is simply not being managed.

Effective Threat Management

A BASIC PRINCIPLE OF MANAGEMENT ACCOUNTABILITY IS TO ask the following question: Who is the most senior person in the organization whose full-time job is dedicated to function x?

Well-managed organizations tend to have clear and simple answers to this question. The most important tasks generally have a full-time manager: for example, finance is handled by the Chief Financial Officer (CFO). The Chief Executive Officer is, of course, interested in his company's finances, but he or she is not working full-time on the finance problem—or on any one problem, for that matter. Similarly, lower-level employees may work full-time in finance, but they are not at the highest level of seniority. The CFO is the highest-level person dedicated completely to the finance function.

Every really important function or problem should have a senior executive dedicated to it, and that executive should have a purposeful organization supporting him or her. A company that does not have a vice president of marketing is not apt to be very marketing oriented. (Whether this is good or bad depends on how important marketing is to the firm's business.) Some problems, by their nature, may take more than one senior-level person working on them full-time; that is fine, as long as there aren't too many cooks and there is a high degree of coordination and cooperation among them.

When one or more senior, full-time people are not dedicated to managing a problem, it almost invariably means that the issue is not important—or that the organization is in trouble. There are many versions of such dysfunction. In some cases, lower-level people are given full-time responsibility but little authority. In others, multiple senior people all claim responsibility for the problem but none of them work on it exclusively. Or the people who do devote all their attention to the problem are scattered among many different departments.

Each of these arrangements is a recipe for disaster. Doing good work under those conditions may be possible, but it is highly unlikely. It is more probable that there will be no cohesive strategy, little ability to implement plans, and no traction. Without clear responsibility, authority, ability to measure performance, and (perhaps most importantly) a structure that holds people accountable, attaining good results is virtually impossible.

Now ask: Who is the most senior government official whose full-time job is defending the United States against strategic terrorism? In the worst possible case, no one is focused on this problem. The lack of a division between strategy and tactics, however, means that even people who are focused exclusively on terrorism are distracted by the enormous volume of low-level threats.

Many public servants, from the president on down, list counteracting some aspect of strategic terrorism as a sideline responsibility. But people who focus on strategic terrorism full-time are relatively low-level government workers employed in different departments and agencies with conflicting missions and no overall direction.

For example, the Strategic Command Center for Combating Weapons of Mass Destruction manages strategic counterterrorism. So do the Defense Threat Reduction Agency and the National Counterterrorism Center—and the Director of National Intelligence, who coordinates efforts by the FBI, CIA, DIA, NRO, DMA, and an alphabet soup of other intelligence agencies. The problem has been dissected into many pieces, which are scattered across the landscape of the U.S. government.

Contrast this situation with our efforts to prevent strategic nuclear war, for which an elaborate and well-defined chain of command exists. We have a comprehensive set of early-warning systems and contingency plans that cover every foreseeable eventuality. Thousands of highly trained military personnel are on duty every hour of every day and are waiting to respond to a strategic nuclear attack.

Strategic Air Command and the Navy's missile submarine command are both part of U.S. strategic defense. Each of these groups, as well as several other components of the nuclear command structure, requires its own leader so that more than one person exists whose full-time job is strategic nuclear defense. An incredibly well-defined set of people have full-time jobs preparing for and responding to a strategic nuclear attack. Their work includes preparing elaborate contingency plans so that the nation cannot be taken by surprise by eliminating part of that chain of command. Our country has taken this task very seriously indeed.

Where are our early warning systems for strategic terrorism? Who is in charge of building them? What is the remedy if an attack takes place? You can't find even the barest outline of a plan, and it is unclear who has the authority and responsibility to draw one up.

Asking who is the highest-level person assigned full-time to a problem is the first question to pose but only the first. Knowing whether the person in charge has a sufficiently powerful organization to support him or her is also crucial. One cannot reasonably expect a top manager or official to coordinate across multiple groups in different departments. This mistake is commonly made both by companies, which sometimes assign senior executives to fix something without also creating capable organizations to assist them, and by governments, which have concocted such ineffectual posts as the president's "drug czar." Titular senior-level posts without direct authority, responsibility, or accountability are next to worthless. Sometimes they even cause harm because their existence lulls us into thinking they are truly providing a solution.

The 9/11 Commission found that the balkanization of the intelligence and law-enforcement communities across multiple agencies and groups partly explains why such attacks could take place without our intervention. The Commission had this to say:

"Surprise, when it happens to a government, is likely to be a complicated, diffuse, bureaucratic thing. It includes neglect of responsibility, but also responsibility so poorly defined or so ambiguously delegated that action gets lost."

The words are not original to them—they first appeared in *Pearl Harbor: Warning and Decision*, and refers to an ex post facto analysis of that debacle which occurred 59 years prior to 9/11. In both cases, poorly apportioned responsibility lead to big, nasty surprises. *The 9/11 Commission Report* continues, "We hope another commission, writing in the future about another attack, does not again find this quotation to be so apt."

Yet without a better organizational structure to manage strategic terrorism, this is exactly what will occur.

The 9/11 Commission focused on the intelligence agencies as those most in need of change, and its recommendations resulted in the establishment of the Office of the Director of National Intelligence. Whether this organizational shift is sufficient to address strategic terrorism cannot yet be determined. Reports to date suggest that the office does not function as well as supporters had predicted. But at least something has been done to unify the intelligence community.

Unfortunately, this logic hasn't been extended to the other state organs responsible for our security. When it comes to devising a response to biological terrorism, no organized community exists to revamp. Is this an issue for the Centers for Disease Control and Prevention? Or should it be handled by the uniformed Public Health Service? Or is the Department of Homeland Security (DHS) supposed to be organizing hospitals? Some elements of each may be required, but at the moment, no cohesion of any kind exists. Token and understaffed efforts are fragmented across dozens of government agencies.

The three monkeys famously convey that that they see no evil, hear no evil, and speak no evil. The bureaucrat's version sounds much the same: "It's not my job; it's not in my budget; it's not my problem."

The passengers on Flight 93 faced a similar issue. Everybody knows you are not supposed to use a cell phone on an airplane—it is against the rules! Passengers are not supposed to interfere with a hijacking. If asked, the airline would surely forbid passengers to take action, if for no other reason than to protect itself from liability lawsuits. Every rule and convention of airline travel told the passengers to stay out of the way, keep their phones off, and not get involved. Don't try this yourself; leave it for the professionals.

The passengers of Flight 93 were smart enough to discard those rules in light of the dire situation facing them. The question for us is whether we will make that same choice as a society. And if we do, will we have enough time for it to matter? Or, like those on Flight 93, will we only reach the decision moments before disaster?

The Sleeping Dogs of War

To understand the government agencies responsible for defending us against terrorism, we must consider the handful of men that influenced the building of American intelligence and defense institutions—men like Adolf Hitler, Hideki Tōjō, Joseph Stalin, Nikita Khrushchev, and Leonid Brezhnev.

As an open, democratic society, the United States is a reluctant warrior at best. We hate going to war and hate preparing for it. We have a very short attention span, and when the crisis of the moment has passed, we lose interest and focus attention and funding elsewhere.

History holds few examples of domestic political leadership spurring effective preparation against strategic threats. The checks and balances built into our system of government so limit U.S. leaders that few, if any, have the mandate or personal charisma required to produce a fundamental change of this kind. When we as a nation do respond, it is typically because of the leadership of an opponent who is persistent enough to shake us out of our lethargy and force us to act. So somewhat perversely, the people who most influenced the construction of our defense establishments were our adversaries.

Before World War II, for example, the United States had turned inward and become isolationist. Only the assault on Pearl Harbor (coupled with dogged scheming by Franklin Delano Roosevelt and egged on by Winston Churchill) persuaded America to confront the threat from Japan and Germany. In a very real sense, Tōjō and Hitler were the fathers of the modern American defense establishment.

Stalin took over where they left off and launched us to the Cold War. This was a long and tiring struggle, but the persistence of the Soviets kept us at it. If at any point American interest or determination flagged, Khrushchev was there to bang his shoe to get our attention. The intense period of competition and rivalry with the Soviets spurred the completion of our defense bureaucracy.

Following the dissolution of the Soviet Union in 1991, no adversary has so dominated our attention. Inertia and the absence of a compelling threat have kept the large bureaucracies in the defense establishment doing largely what they had done before.

The 9/11 attacks seemed to herald a new era because the shock they inflicted was as great as any since Pearl Harbor. Unfortunately, 9/11 alone was not sufficient to set the nation on a new course. Instead, when the alarm went off, we startled awake briefly but then hit the snooze button and fell asleep again. Even Al Qaeda has not inspired the kind of deep organizational change needed to shake our defense and intelligence establishments out of their post–Cold War slumber. The 9/11 attacks and subsequent military operations in Afghanistan and Iraq have brought some changes. But the vast machinery of the Cold War, built up over five decades, has yet to retool.

To the extent that alterations have been made, they are primarily the most obvious ones—for example, correcting the egregious problems in immigration and airport security that contributed to 9/11. No stroke of genius was required to recognize the problems in the Immigration and Naturalization Service after it approved visa extensions for some of the 9/11 hijackers—a year after they died in the attacks. That kind of gross incompetence has been noted and, one would hope, corrected. The INS has since been split up and its responsibilities spread across several federal agencies, hopefully for the better.

The 9/11 Commission hearings highlighted similar problems with U.S. intelligence before the attacks, and this lead to similar reshuffling of authority including the creation of the post of Director of National Intelligence. Perhaps this too will help, but the commission's report focused on events that took place on 9/11 and before.

If our future threats were always the same as those of the past, the Commission's recommendations might be adequate. Unfortunately, there is every reason to believe that the most significant dangers we will face in coming years will be completely new ones. Thus, the precautions we take must be novel as well. The steps necessary to prevent nuclear and biological terrorism are qualitatively different from those needed to plug the holes that allowed 9/11 to happen. Yet our military forces and government agencies seem not to recognize this difference. Nearly all of their personnel and resources are focused on the immediate problems posed by tactical issues in Afghanistan and by low-level terrorism directed at the United States.

Immediately after 9/11, the attention was there, but it has largely dissipated. We have reentered a politics-as-usual period in which making any fundamental changes is difficult. The challenges posed by short-term problems outcompete new issues like strategic terrorism, and the situation is not viewed as dire enough to make the country willing to get really serious and plan for the future.

The chief reason for our complacency is that no major attacks have occurred on American soil in the last 12 years. Suppose that one 9/11-scale assault on American cities had occurred every six months to a year since 2001. Large buildings regularly exploding in Los Angeles, or Chicago, or Boston would be impossible to ignore. By now, we would be getting pretty damn serious about the problem—both in the short term and in the long run. That kind of threat spurs an otherwise complacent country to action. Without it, the urgency necessary to cut through layers of bureaucracy simply does not exist.

Instead, our passive vulnerability is likely to grow over the next few years until another attack occurs. Al Qaeda and the other enemies of America are not like Hitler, Tōjō, or Khrushchev—they will not confront us directly. Instead, they will wait patiently for an opportunity that allows them to strike hard with stealth.

The Long View, Backward and Forward

YOUR CAR HAS A VERY LARGE WINDSHIELD, THROUGH WHICH you can see the road ahead, but only a few small mirrors to view what is coming up behind. That's because the threat is largely from the front, from the direction in which you are moving.

A bureaucracy (particularly one that exists within a democracy) has the opposite arrangement: an enormous rearview mirror and just a tiny peephole facing forward. The structures and mandates of bureaucracies are based on what has already happened, not what *will* happen. They cite history to justify their operations. Actions based on a view into the future are speculative and open to criticism, especially when the problems of the present loom large. The only force that has a proved ability to shake the complacency inherent in bureaucracies is a determined adversary that persistently fights us or antagonizes us. Yet here, Al Qaeda is unlikely to indulge us.

Instead, Al Qaeda is more likely to resemble Ho Chi Minh. No large agency owes its existence to Ho Chi Minh because the U.S. never figured out how to fight his Vietcong guerillas effectively. In the contests with Nazi Germany, Imperial Japan, and the Soviet Union, we were pitting our bureaucracy against theirs, and ours was superior. In Vietnam, our bureaucracy confronted a determined foe fighting a long-term guerilla war on its own soil. To the U.S. defense establishment, that wasn't a fair fight, so we lost. This perception certainly left its mark on our nation's military, but it was not a commitment to build something new. Instead, it shook our confidence and taught us to think of military intervention as a series of misadventures by the gang who couldn't shoot straight. If Al Qaeda has the patience of Ho Chi Minh, it will time the interval between major attacks to be just a bit longer than our attention span, ensuring that Americans will always be asleep at the switch when the next assault occurs.

For much of the last decade both we and Al Qaeda have been distracted by Iraq and Afghanistan. The growing malcontent in the United States about the war in Afghanistan is obvious, and the continued loss of American life in small-scale skirmishes and terrorist attacks there accounts for much of the angst. Our adversaries are smart, and they have concentrated their attacks on Afghanistan and Pakistan to further undermine public support for the war and to put pressure on the United States to pull out, which we intend to do by the end of 2014. Whether that will result in a lasting peace and a stable Afghanistan, or play right into the hands of our adversaries remains to be seen.

This concentration of terror attacks is arguably good for the U.S. public in the sense that our soldiers in Afghanistan and our drones in Pakistan are better equipped to handle the challenge than we are here. It's also good for Al Qaeda—especially since U.S. military and intelligence efforts have disrupted their organization—because attacks in Afghanistan are cheaper and easier to mount than more direct operations against the United States would be. Attacking the U.S. mainland now would only antagonize the American public and recommit us to the war on terrorism.

Eventually, this strategic calculus will change. Whether it changes tomorrow or in 2033 is impossible to predict, but it is hard to believe that another major attack won't occur within a generation. If the next major incident is only a 9/11-scale, M3.5 attack, it will be traumatic, but our society will survive largely intact. The problem is that we are not apt to be that lucky.

The clear pattern of Al Qaeda—from Somalia, to Khobar Towers, to the African embassy bombings, to the U.S.S. *Cole,* to the World Trade Center and the Pentagon—is one of infrequent attacks, the worst of which escalate in severity. The next one could be an M5.0 or M6.0 nuclear or biological event. Waiting until it occurs to begin our preparations is utterly irresponsible, but that is just what we're doing.

Meanwhile, we are only beginning to look beyond Al Qaeda to the motives and capabilities of those groups that will one day succeed this faltering foe. This is alarming when one considers that 20 years ago, neither Al Qaeda nor any other radical Islamic organization would have made anybody's list of major threats to U.S. security, a list which at the time was dominated by the communist superpowers. Back in the 1980s, we were funding our *ally* Osama bin Laden in a guerilla war against the Soviet occupation of Afghanistan as a minor part of our decades-long Cold War with international communism.

How ironic that today the communists are gone, save for Kim Jong Un's North Korea and Raúl Castro's backward little island. Kim and Castro are the proverbial exceptions that prove the rule, the last men standing of the communist strongmen. The successor states to all of the other communist states are now thoroughly capitalist and are large trading partners of the United States. Vietnam is a thriving, market-driven economy and, I am told, a great place to vacation. China launched, in our direction, millions of tons of hardware—not weapons, but store loads of Wal-Mart merchandise. Meanwhile, our erstwhile ally bin Laden became public enemy number one.

Twenty years from now, new terrorist groups and causes will exist. Radical Islam is likely to remain a concern in 2033, but it won't be the only one. Some of today's players will leave the international arena, and new ones will enter. But strategic terrorism is here to stay.

In future decades, the United States may not even be a primary target of terrorists. But that notion brings little comfort, given the dangers of biological terrorism: smallpox released in Moscow by a Chechen terror group or in Beijing by Uighur separatists from China's western provinces could infect the United States just as easily as a New York City–based attack could spread to Russia or China.

If a natural epidemic like SARS can shut down Toronto hospitals within days of emerging in Asia, surely the same will be true of deliberate attacks. Any small group seeking a terrorist resolution of its grievances could kill us as part of the collateral damage.

It is crucial that we realize that the fundamental problem is not limited to a specific organization like Al Qaeda or to a specific ideology like radical Islam. Bin Laden is dead and gone, yet this general threat persists. Just as managing nuclear weapons became a permanent part of the world order after World War II, combating strategic terrorism must become a permanent part of ensuring global security today. This challenge demands some dramatic shifts in American defense and foreign policy. It isn't a temporary crisis, and it requires a fundamental and long-lasting adjustment to the new state of affairs. The investment needed is similar in scale to that spent during the Cold War—hundreds of billions of dollars, perhaps more. To pay for this, difficult decisions will have to be made at many levels.

One source of funding may be obtained by purging the anachronistic elements of our defense establishment, those that are still fighting the Cold War. But reallocating such spending is likely to be insufficient because of the magnitude of the task and the difficulty of prying funds loose from existing programs. We will need to tighten our belts in ways that we have not done for decades. The Cold War influenced all aspects of American foreign and defense policy (and many aspects of domestic policy as well), and the war on strategic terrorism, if it is to be effective, must have similar scope and impact.

That doesn't mean a Cold War revival because there are deep differences between then and now. The Cold War was about building a deterrent—implementing the strategy of mutually assured destruction for any party foolish enough to initiate nuclear hostilities. As such, it was relatively straightforward: create a defensive deterrent by building ever more terrible offensive weapons in multiple redundant systems.

The war on terror is fundamentally different. We cannot win by developing more powerful offensive weapons than our adversaries.

Deterrence of the old sort simply does not work. New weapons are needed for a portion of the struggle, but only for a portion. The task is nowhere near as simple as building bigger, more, or different kinds of weapons, as it was during the Cold War.

Another difference between the Cold War and the war on terror is the growing irrelevance of nation-states. The world's laws—and our thinking—remain oriented around an international system based on nation-states, but this system often impedes effective action.

The nation-state is the fundamental unit of international diplomacy, law enforcement, and discourse. We assume that a country is responsible for its sovereign territory. When a criminal crosses a national border, we rely on the country he or she then resides in to handle the arrest, and we go through a formal extradition process to get that nation to hand over that criminal. This hierarchical approach is rendered useless when a tiny group can create weapons that threaten the population of entire continents.

During the last 20 years, terrorists have exploited the general respect for national boundaries by seeking the shelter of countries that either have actively supported them or have looked the other way. Terrorist training camps have been located in places such as Libya, which fostered terrorism, or in lawless zones such as Somalia, Sudan, and Afghanistan.

Yet the 9/11 plot was not implemented as a direct attack from Afghanistan. Training camps in that country may have helped prepare and inspire the men who commandeered the planes, but the plot was conceived initially in Germany, and all of its leaders did their work in Western Europe or the United States.

Strategic terrorists have even less need for training camps. A strategic terror attack, whether nuclear or biological, will most likely be planned by people in Western Europe or the United States. Indeed, they will be safest there because laws in these countries protect individual freedoms. Terrorists in a desert outpost in Sudan or a cave on the Pakistani tribal frontier have to worry about Special Forces commandos, Tomahawk cruise missiles, and Hellfire rockets fired by robotic Predator spy planes. In Paris, Munich, or San Diego, they won't have any such concerns.

What, then, do we do when a terror group hides within a friendly democracy? Do we count on authorities there to police the territory inside their borders? Do we hold them responsible when they fail?

Aum Shinrikyo provides an instructive case in point. Members of this cult produced their biological and chemical weapons at facilities in Japan and did experiments on a remote ranch in Australia. They went undetected at the time by officials in either country. When Japanese prosecutors finally charged them after the attacks, it was for a grab bag of random violations. Japan had no laws against the creation of biological- or chemical-warfare agents. Why should they have? Such agents had probably never before been made by civilian groups, and certainly not in Japan.

Aum Shinrikyo clearly wanted to target Japan, but what if its leaders had planned to unleash their anthrax and sarin elsewhere—say, in the United States? Surely a responsible, law-abiding country such as Japan would have wanted to cooperate in stopping such horrors. But without laws against preparing such weapons, authorities may have found interdiction difficult. How constrained by Japanese laws and sovereignty should the United States be in finding such agents and in stopping such forms of attack?

If that issue seems simple, change the group and setting. What if Chechens based in, say, Boston were working to damage Russia but were violating no U.S. laws. Would we want Russian security agents to run amok in Beantown? Probably not—we take our sovereignty and legal rights very seriously. Terrorists are much better at exploiting these loopholes in the structure of international law than nation-states are at closing them.

Indeed, the passage above about the hypothetical case of Chechen terrorists in Boston was written in an earlier draft of this paper. Subsequently, actual Chechens perpetrated the very real Boston Marathon bombing.

Although the target turned out to be the U.S. rather than Russia, clumsy handling between nations was a factor. It turns out that Russian security agents did warn us about the Tsarnaev brothers, and U.S. officials gave them only a perfunctory look before telling the Russians everything was fine. This apparently included missing the fact that one of the brothers was involved in multiple Boston area murders.

Clearly we need new international tools to combat this threat. The aftermath of World War II and the Cold War created a number of new international groups and structures—including: the United Nations (UN), the North Atlantic treaty Organization (NATO), and the Warsaw Pact. These international groups were a direct reaction to the challenge posed by the threat of nuclear war and the emergence of highly polarized communist and capitalist ideologies.

We must develop a new set of extra-national organizations, perhaps akin to NATO or UN, to cope with the new threat of strategic terrorism. Convincing our former rivals of this need should not be difficult, at least in principle. Indeed, Russia and China have far more common cause with the United States in this regard than they do with protégés such as Cuba or North Korea.

Reaching out to traditional enemies is a difficult process, one that requires time to establish trust. Reworking the international framework of nation-states is an even tougher task because a tremendous amount of history and entrenched bureaucracy lie behind it. But until the hide-bound international community struggles to reorient itself, strategic terrorists will have opportunities to operate decisively and with impunity in many places.

The Audacity of Courage

Studying this issue without becoming depressed about our prospects is challenging. A free and open society is hard to defend, and the insane lethality of current and future technology makes it even more difficult. In contrast, believing that all efforts will be in vain is easy.

My friend Sir Martin Rees, a distinguished professor at the University of Cambridge and Britain's Astronomer Royal, is so pessimistic that he believes no hope exists for the human race. He published a book to that effect in 2003 called *Our Final Hour*. In it, he argues that somebody will unleash an M10 event and drive our species to extinction or, at the very least, will end civilization, and he sees guarding against that as futile. His argument is hard to discount.

Sir Martin may be right, but giving up seems foolish. All life is a struggle against inevitable death, and yet each of us obtains all that we hold dear during that struggle. We can forestall and, with luck, prevent total catastrophe in the long run. And in the meantime, we can greatly reduce the likelihood and severity of smaller terrorist attacks. This problem is not impossible to solve, just very difficult.

Global thermonuclear war with the Soviet Union was once considered equally unavoidable, and thinking that we were all doomed was once fashionable. As children, my classmates and I did "duck and cover" air-raid drills to prepare for what seemed then like the inevitable apocalypse. At various points in the 1950s through the 1970s, brinksmanship by the United States and the Soviet Union nearly did trigger a nuclear exchange, but we avoided it. It was not easy, and the actions taken were certainly not free of risk, but the world survived despite many dire predictions.

Indeed, those predictions were an important part of the solution. When discussing bioterrorism, I'm frequently told that I'm being a scare-monger, just like those who exaggerated the danger of global nuclear war, which didn't happen. But if nuclear weapons had not been widely feared, would all those actions that have been taken to avoid their use been done? The paranoia of military planners and that of antinuclear activists were both important. Nuclear weapons truly scared people, hawks and doves, Soviets and Americans alike. But frightening people by itself isn't enough. Instead, fear has to be mixed with something more actionable—a plan to allay the fear.

The problems of strategic terrorism can be solved, but the solution will take more than spending $1 billion here or there. It will be a multi-decade struggle that will affect as many aspects of people's lives as the Cold War. Indeed, the Cold War provides us with a useful model. One can list the ways in which the Cold War affected scientific research, intelligence gathering, military planning, the practice of diplomacy, public policy making, and other activities. In the same manner, one can also prospectively lay out what we need to change in each of these spheres of action. Each has a special set of challenges that must be met.

The Research Challenge

In most wars, scientific research is a secondary activity rather than a frontline effort. This approach is emphatically not appropriate in the struggle against strategic terrorism, which is primarily a technological and scientific battle. Unfortunately, we are at least 20 years behind our adversaries. For more than two decades, we have allowed an unprecedented explosion of work in molecular biology to occur without providing any substantial funding for understanding and preventing the misuse of this knowledge and technology.

Scientists routinely publish results that either implicitly or, in many cases, explicitly contain recipes for mayhem. Many such findings have already been openly disseminated worldwide in the scientific literature. Broadcasting such knowledge could have incredibly lethal consequences, but no funding agency has yet devoted substantial resources to figuring out what the threats are in detail and to developing countermeasures against them. Ironically, this sort of research is precisely the kind at which our society excels. But developing such solutions will be impossible if we don't bother to identify the problems and do the work. As it stands, we don't. The reason is simple enough—little, if any, funding is available for countermeasures research. The National Institutes of Health (NIH), the National Science Foundation, and other government grant-making agencies provide research funding in biology and medicine, but their priorities are arranged to combat natural scourges, not bioterrorism. Why? Because plenty of here-and-now diseases, such as cancer, diabetes, and AIDS, are vying for their attention. Spending money to fight speculative future threats is far more risky and, hence, is rarely done.

Part of the problem is that politicians tend to respond to strenuous lobbying. As a result, research has increasingly focused on certain diseases at the expense of others. HIV/AIDS is the great winner here, superseding diseases that kill far more Americans. If you divide total NIH funding by the number of Americans affected each year by the disease, you find that HIV/AIDS gets 10 times more funding on a per-patient basis than does breast cancer, 50 times more funding than diabetes, and 100 times more than heart disease. If heart disease, cancer, and diabetes kill far more people than HIV/AIDS, why is research funding so skewed? In part, it is because the study of HIV/AIDS encompasses many fundamental biological questions, so it is a rich area for research. Scientists like to research HIV/AIDS because they get to look at many basic scientific questions.

In part, it is because of fear that AIDS could worsen in the future and become a larger epidemic in the United States (as it has in some other parts of the world). Another factor is that HIV was novel when it emerged; cancer, heart disease, and diabetes are all old hat so we've become inured to them. But much of the reason is that AIDS activists have been very effective in lobbying Congress to support AIDS research.

Bioterrorism has, so far, killed only five people in the United States, so one cannot make a case for greater funding based on past mortality. But basing these decisions on historical precedents is another example of driving by looking in the rearview mirror.

It is ironic that government sponsors of some of the largest biomedical research programs use war as a metaphor. For example, the Nixon administration launched the "war on cancer," and since then, we've declared "war" on AIDS and on many other diseases. Yet when it comes to strategic terrorism, we face the prospect of a real war but have yet to get serious about doing the needed biomedical research. The only way to change this situation is to forge a comprehensive plan for research, development, and deployment of technologies to detect, cure, or prevent a biological attack. (Similar plans need to address nuclear and other possible varieties of strategic terrorism, although we will most likely be able to draw on existing work more directly to thwart those threats.)

Such a plan could be implemented quite rapidly. Defining the research objectives is straightforward: it would include both directed research aimed at specific problems and risky exploratory work.

A program of this kind would have many follow-on benefits. Such benefits resulted during the Cold War when the military and its Defense Advanced Research Projects Agency (DARPA) sponsored virtually all of the important academic work in computer science. The result was a technology boom that has vastly enriched the world during the last 30 years.

A well-funded research initiative to develop bioterrorism defenses would give an enormous boost to biomedical research in some areas that may ultimately prove just as useful. Such a move would also do a tremendous amount of good for other parts of the interconnected world in which we live. At the moment, all of humanity is susceptible to natural infections that are very similar in some ways to those that might be unleashed during an act of bioterrorism: a novel strain of pandemic influenza, say, or an emergent pathogen such as the one that causes SARS. Counter-bioterrorism research could lead quite directly to broad-spectrum antiviral drugs and vaccines or to monitoring systems for detecting outbreaks early. We could expect enormous dividends from this research in areas well outside of bioterrorism defense itself.

Scientists will rise to this challenge if given adequate time and money—indeed, the United States excels at such scientific and technological research. Still, considerable patience will be required: countering strategic terrorism isn't a single, isolated problem like building the atomic bomb. It is not a Manhattan Project, with a single, focused, technological objective—instead, it is 100 disparate problems.

Any discussion of government-funded research must deal with the ghost of the Manhattan Project—the fantastically successful effort to build the first atomic bomb. The Manhattan Project was a nearly miraculous achievement by any standard. It is fondly remembered as a time when a fairly small core group of scientists in the high desert in Los Alamos, New Mexico, accomplished the seemingly impossible tasks of inventing enough nuclear physics to conceive various bomb designs and of figuring out the key technical steps to build the atomic bomb. Their work ended the war with Japan, which was otherwise estimated to require the loss of a million American lives and many times that number of Japanese ones. At the same time, it gave us the horror of nuclear weapons. But no matter how horrified one may be at the product, the project's success in achieving its stated goals cannot be denied.

The Manhattan Project is extremely compelling as a model because it seems to offer both small size (just a few nerds in the desert) and miraculous results (a way to end a brutal war quickly). Every time a technical challenge appears, somebody in government will say, "Let's get our best minds working on this problem," and invoke the spirit of the Manhattan Project as inspiration, either implicitly or explicitly.

Other examples exist of small scientific teams making gigantic contributions, especially during wartime, when bureaucratic rules do not apply and the very smartest people are willing to pitch in. The British effort at Bletchley Park to decipher the German Enigma code machine is one such instance. Without Enigma decryption, many historians argue, Britain would have been forced to surrender to Hitler long before the United States entered the war.

These famous examples of the products of small teams of brilliant people are very tempting to try to emulate. When you desperately need a rabbit pulled out of a hat, resisting the draw of a magical solution is hard.

The alluring notion that a small crash project can save the day gets lots of attention—unfortunately, more than it deserves.

Looking to some romantic view of the Manhattan Project as a model is wrongheaded, for several reasons. First, the project to create the bomb was no minor effort. It ultimately employed an enormous number of people all over the United States. Second, strategic terrorism is not a single, isolated task like building an atomic weapon; it comprises 100 different tasks. The Manhattan Project created just two nuclear weapons that won the war, but bioterrorism requires a fundamentally more complex solution.

Another myth of the Manhattan Project is that it was small—a handful of supersmart scientists sequestered at a secret desert lab. While the team at Los Alamos was indeed that, at least in part, the overall scope of the project was enormous and spanned facilities across the country. The mythology that has grown up around the Manhattan Project makes it seem simpler and more affordable than it really was.

Finally, the analogy to an idealized Manhattan Project tends to let policy makers off too easily. It is much cheaper to convene a small team than a large one and to do a crash program rather than patient long-term research programs. Indeed, once you start believing that a crash program in the 11th hour will work, it saps the will to fund the longer, more painstaking work. We may need to mount Manhattan Project–scale efforts for some of the key bioterror problems, but overall, the analogy is entirely inappropriate.

The situation is not hopeless, however. Many promising avenues of research could be followed. The R&D capabilities of the United States are still unmatched in the world. A full description of the research agenda is beyond the scope of this treatment, but it could be put together in short order.

The Intelligence Challenge

PREVENTING NUCLEAR WAR AND FIGHTING COMMON CRIME ARE similar in some ways. Both efforts typically exploit the principle of deterrence by inflicting punishment after the fact. This approach works well when the deterrence is real—that is, when it is clear that the probability of punishment or retaliation is high. In the case of strategic terrorism, deterrence doesn't work for the simple reason that we cannot retaliate effectively.

Besides deterrence, the other main approach to security is guarding: preventing crime by having forces on the scene that stop criminals or attackers in their tracks. Guarding is used quite a bit in counterterrorism—that is what air marshals on flights and security screeners in airports are doing.

Unfortunately, guarding does not prevent strategic terrorism. If the goal of our terrorist adversaries is to spread an infectious disease in the United States, the simplest way to do it may be to put a few infected volunteers (who need not even know they are infected) on a plane headed into our country. It would be difficult for security screeners or air marshals to notice anything amiss: the terrorists wouldn't be obviously sick, and they wouldn't be carrying guns, knives, or other suspicious items. Even if a way existed to detect such attackers, by the time someone found them in the United States, it would already be too late.

Contagious bioterrorism presents perhaps the worst case, but even with nuclear, chemical, or noncontagious biowarfare, guarding the country is of limited use. Intercepting a nuclear bomb in a shipping container works only if you stop it in a place you don't mind losing if the weapon detonates upon discovery. In principle, having a nuclear bomb explode in a Port Authority facility in New Jersey may be marginally better than having it explode in midtown Manhattan, but it would be a Pyrrhic victory.

The only way to beat strategic terrorists is to go after them, either in their home territory or, if they are already here, before they have built a weapon that is sufficiently dangerous. You need to strike preemptively.

The Iraq War, however, has given preemption a bad name. A major public justification for the war was to destroy Saddam Hussein's weapons of mass destruction, yet investigators ultimately found that he didn't have any. This realization not only indicts the intelligence process that led to that estimate; it also discredits preemption itself. The result is that both the country and the world will be highly skeptical of any new rush to a preemptive attack. Most preemptive action will not be at the level of a full-scale war and thus will require lower thresholds of certainty. Nevertheless, any sort of preemptive attack places tremendous demands on intelligence gathering, demands that our intelligence community, in its current form, cannot meet.

At the highest level, the need to battle strategic terrorists preemptively sets the bar for 21st-century intelligence services: they must provide information of sufficient quality and timeliness to enable policy makers to decide whether or not to act. If we can build an intelligence community that does that, we will be in good shape. If we can't rise to that challenge, either because we don't know enough or because we don't have enough confidence in the information we have, then we will fail—which is exactly what we concluded after 9/11.

The 9/11 Commission produced a 565-page report outlining various shortcomings in the intelligence leading up to the 9/11 attack, including numerous missed opportunities to arrest or intercept the hijackers or otherwise foil the plot. It also made broad recommendations about how to change the U.S. intelligence community to prevent the systematic lack of coordination that caused the community to miss the 9/11 attack.

The specific 9/11 Commission recommendations are mostly beside the point. The clear objective of the Commission was to prevent the specific errors that allowed one incident of tactical terrorism to occur. The goal was to lock the barn after the horse got out. Its focus, like the 9/11 attack itself, was almost entirely tactical in nature: No consideration of strategic terrorism is offered in the report. Instead, it concerns itself with improving communications and coordination within the existing intelligence paradigm.

But the paradigm itself must change. Too much of our intelligence infrastructure was built for the Cold War or for a world in which information technology was largely static or slow moving. Changing that mind-set will, in some cases, demand a fundamental rebuilding of agencies whose purpose, process, and technical assets have changed little since the Cold War. (Indeed, the failure of the U.S. intelligence community to correctly assess the state of Iraq's programs to develop weapons of mass destruction demonstrates that these institutions have not even been able to carry out effective surveillance of traditional nation-states.)

In the arcane terminology of the intelligence community, signals intelligence (such as determining the properties of a radar system) is called SIGINT. Communications intelligence is COMINT, and good old-fashioned human spying is HUMINT. There are many more. Each of the "INTs" needs a complete bottom-up review to determine whether its structure and methodologies match present and future needs. The question to ask is simple: Can we count on that intelligence to give us clear and convincing warnings that we can act on?

Revamping our intelligence services won't be easy. The new approach will require large budget increases, which will not be popular with politicians or the public. An obvious place to look for the money will be the budgets of existing programs, which will be defended by members of Congress in whose districts they provide employment. The contractors will howl, and the bureaucrats will put up roadblocks.

Action is nevertheless imperative. Perhaps the largest and hardest step to take is the separation of strategic from tactical terrorism. Without a clear separation, the right spotlight cannot be put on the problem. Yet this division will also mean shifting resources away from tactical terrorism. In the case of the DHS, this step would be controversial, to say the least. The DHS has focused almost entirely on tactical terrorism.

The Military Challenge

GATHERING INTELLIGENCE IS ONLY THE FIRST STEP; THE second is what to do with it when it indicates a threat exists. As I've argued in previous paragraphs here, the only really effective approach is preemptive intervention. Unfortunately, calling for such intervention places enormous demands on intelligence gathering and even greater demands on the leaders who decide to intervene. What is the threshold for action? What sort of team do you send in? What are the risks of collateral damage? What if we're wrong?

The primary military challenge in addressing strategic terrorism is to develop enough depth and breadth in new forms of special operations to give decision makers an appropriate set of options. Taking out a terrorist camp that is building a nuclear weapon or brewing up smallpox is a very specific challenge. What if that camp is in a city? What if it is in an American city? The years since 9/11 have seen an increase in the size and importance of special operations, but this increase appears to be a small down payment on the capabilities the future will demand.

We need to develop new weapons, for example. At the moment, our military lacks practical weapons that can destroy a bioweapons facility in a way that guarantees the contents are sterilized. Although the so-called "surgical" air strikes of the past have improved greatly, a tremendous amount of collateral damage still occurs. Weapon systems must be rethought and optimized for a wide range of special operations, from small-scale covert action to large-scale efforts such as the current one in Afghanistan.

The U.S. military also needs to adjust parts of its organization. Troops involved in special operations—Rangers, Green Berets, Delta Force members, and so forth—have long been treated as adjuncts to the "real" forces. This attitude is another holdover from World War II, when commandos did the prep work for conventional assaults. Such an approach is unlikely to be useful in most future engagements. Instead, attacks will often use special-operations units without involving any conventional forces. It may even make sense to unify all special operations under a separate branch of the armed services, one more on par with the Army, Navy, Marines, and Air Force than SOCOM is today.

Another major organizational shift would be to extend the reach of military operations inside the United States. Most acts of terrorism are treated as criminal activities and are thus left to police SWAT teams or the FBI. But it is unrealistic to expect such forces to have the specialized training and expertise to deal with strategic terrorist threats. Instead, we need a nationally trained, and nationally funded force, even more developed than the FBI's WMD Directorate and Critical Incident Response Group (CIRG).

Suppose that intelligence efforts do locate a nuclear weapon but only after it has entered the United States. Who goes to get it? It seems rather unlikely that the local police or SWAT team would be prepared to handle it. The FBI has a special hostage rescue squad, but what if an air strike is required? The FBI doesn't have fighter jets, although it recently revealed a limited use of drones, which generated some controversy.

The United States has a long history of separating military power from civil authority and an equally long history of dealing with invasions, which clearly require military power. Strategic terrorism is a challenge to our preconceived notions because it spans the full range of possibilities from a criminal act by one or a small number of people to an all-out invasion. This is only one of many urgent challenges to military policy.

The Domestic-Policy Challenge

One of the most difficult challenges arises from the clash between the American tradition of liberal freedoms for its citizens and the extreme circumstances of strategic terrorism. We take it for granted that people should enjoy a wide array of rights. The most extreme version of this sentiment was expressed by Sir William Blackstone, an 18th-century jurist: "It is better that ten guilty persons escape than that one innocent suffer." There is substantial practical logic to the idea that trampling on the rights of millions of free citizens is, in aggregate, worse than letting a small number of criminals escape justice. The implicit calculus of harm is that whatever havoc a guilty party may wreak is less odious to society than the damage that may be caused by prosecuting the innocent or abridging their rights through unreasonable search and seizure or other proscribed police behavior.

One way to view this clash is as a trade-off. On one side are the collective rights of the innocent citizenry; on the other are the rights of the victims of crimes that could have been avoided. When a court releases a criminal on a technicality or prevents the police from doing a search or placing a wiretap, it creates a situation in which a future victim could be hurt. How many people are we willing to sacrifice in this way to maintain the civil liberties of the populace? Probably quite a few.

Putting it this way may sound extreme, but we Americans sacrificed around 30,000 people in highway deaths in 2011 alone in exchange for our freedom to drive. Asking for sacrifices in return for civil rights is not untoward. Indeed, with respect to lives lost to criminal activity, civil rights appear pretty cheap: slightly more than 14,600 murders occurred in 2011. Knowing how many of these deaths could have been prevented through repeal of the Bill of Rights is impossible, but the number is unlikely to be more than a small fraction, perhaps a few percent at most. Surprising as it may seem, the unwritten "right" to drive is expensive in terms of the cost in lives, whereas our civil rights are a bargain.

This difference arises because relatively few criminals exist and the amount of harm each of them can do is relatively limited. Every now and then, a monstrous serial killer such as Ted Bundy surfaces, but the number of criminals who have taken more than a handful of lives during their entire career is tiny indeed. Police could doubtless do a better job if they didn't have to respect our constitutional protections, but career criminals are usually caught anyway. They'd just be apprehended sooner or suffer worse penalties if the police had more access to information. The incremental cost to society imposed by the Bill of Rights and other liberal freedoms is thus the product of a small number of lawbreakers, the limited amount of damage these criminals do, and the modest degree by which the conviction rate would improve without safeguards for our civil rights.

When the Founding Fathers established the Constitution and the Bill of Rights, constraining the power of the state in this way was a radical, untested, and unprecedented experiment. Fortunately, the experiment worked! Liberal protection of human rights, pioneered on a large scale first in the United States and then exported to Europe and other developed nations, has been a great success.

Indeed, the project has been expanded substantially. The actions of the U.S. Congress in writing new laws, of the courts in interpreting the Constitution, and of advocacy groups such as the American Civil Liberties Union and a very active criminal-defense bar have expanded the civil rights of Americans. One can hardly credit Sir Blackstone or the Founding Fathers for protecting us from government wiretapping—it didn't exist in their time. Nor did the police read 18th-century suspects their Miranda rights. Prohibition against racial profiling is another modern example—in Sir Blackstone's era, slavery was legal, and discrimination the norm. The number of rights and privileges enjoyed by Americans has steadily increased over time. And the experiment has continued to work.

This entire endeavor is, however, called into question by the nature of strategic terrorism, whose potential for harm is enormous enough to demand a reexamination of the quantitative bargain. Is the cost to society in lives really worth more than the cost of constraints on civil liberties?

Sir Blackstone's trade-off implicitly assumes that the harm done by causing one innocent man to suffer is worse than whatever harm the 10 guilty men may do with their freedom. Let us accept that he was correct in the 18th century or even in the 20th century. Is he still correct in the 21st century if those 10 guilty men are strategic terrorists who could kill millions of innocent Americans?

An innocent man convicted is perhaps the most extreme example. Nobody wants to see that, in part because of our natural sense of compassion, and in part because each innocent person thinks "but what if it was me?" But, even if you accept Sir Blackstone's premise about erroneous punishment, do the other, more recently recognized rights make sense? As a recent example, the ongoing scandal started by Edward Snowden hinges, apparently, on the belief that we all have a right to keep phone records of calls to foreign terror groups private from the NSA. How does the harm versus good calculation work out in that case?

The argument made by civil libertarians who defend Snowden appears to be that (in Sir Blackstone's format), "it is better to let 10 terrorists operate inside the United States than have the NSA see our Verizon bill".

The purely principled tend to reject quantitative arguments—instead, they regard civil liberties as absolutes that must not be subjected to a cost-benefit equation. But, morally speaking, can that really be true? Can those charged with protecting public safety really make absolute tradeoffs without considering that millions of deaths could result?

Civil-rights advocates say yes. They contend that legal precedents indicate that once you allow some backsliding, you step onto the slippery slope and slide into a neofascist police state. In that case, the cure may be worse than the disease.

These are very serious issues that need to be weighed carefully and rationally, but Americans tend to overreact from emotion and lurch from one extreme to another. During peacetime, we expand rights steadily.

Yet when the chips are down, we have routinely violated those rights in ways that were not simply unconstitutional but also ineffective and most likely unnecessary. One of the most shameful examples was the internment of Japanese-Americans during World War II. This program was both deeply racist (German-Americans and Italian-Americans were never interned en masse) and also of dubious efficacy. The modern version of this debate focuses on the Patriot Act, the definition of torture in military and CIA interrogations, the detainment of "enemy combatants" at Guantanamo Bay, or the scandals about warrantless wiretapping and e-mail interceptions within the United States by the NSA and other such government agencies.

Ham-fisted, inept, incompetent, or flatly illegal government actions, like the deplorable treatment of Iraqi prisoners at Abu Ghraib prison, cast a long shadow and reinforce the point of those who seek to defend our civil liberties. The set of rights we are willing to surrender to the government is, in large part, a function of how competent and fair we think the government will be. Unfortunately, ample evidence exists that the government can be unworthy of our trust.

In addition to the challenges inherent in addressing these difficult issues, the debate is often so highly polarized by political factors that rational discussion becomes difficult. Thus, the domestic-policy challenge poses a key question: How can we tackle the issue while rationally steering a balanced course between protecting citizens from misuse of government power and protecting them from strategic terrorism?

The Diplomatic Challenge

Whatever trade-offs we make in our own country, the approach we take to strategic terrorism is also important in the international setting. How can we create a set of diplomatic relationships to help stop strategic terrorism?

At first blush, the answer to this question may seem straightforward—cooperation in addressing nuclear or biological terrorism seems to make sense for all concerned. But it is never quite that simple. Most countries will continue to have conflicting parochial interests. This situation can be seen very clearly in the international debate about the wars in Iraq and in Afghanistan. Political, diplomatic, and other forms of cooperation exist today and include some cooperation at the intelligence and counterterrorism level. But if we are to succeed against strategic terrorism in the long run, much more cooperation is required.

NATO is an interesting example of a diplomatic structure created by the need to address concerns about Soviet expansion and nuclear war. NATO and its mirror-image group, the Warsaw Pact, broached issues that were previously held to be the prerogatives of individual nations. Getting the military of one nation to cooperate that closely with those of other nations was not easy, yet this is exactly what happened.

By analogy, the prescription for strategic terrorism would be to extend NATO-like cooperation from the military to intelligence and counterterrorism organizations. With the set of nations currently in NATO, this approach may be possible. To be really effective, however, future strategic-terrorism alliances would have to include Russia and China as well as some Islamic countries. Is such cooperation possible and practical?

NATO would have been impossible without the background of World War II and the perception of a shared Soviet threat. It is far from clear that sufficient will now exists to mount a similar effort to battle terrorism. After a few million deaths in multiple countries from a bioterrorism attack, we would undoubtedly find the will and the way. So the question is: How much can we accomplish *before* such a tragedy occurs?

Conclusion

Several powerful trends have aligned to profoundly change the way that the world works. Technology now allows stateless groups to organize, recruit, and fund themselves in an unprecedented fashion. That, coupled with the extreme difficulty of finding and punishing a stateless group, means that stateless groups are positioned to be lead players on the world stage. They may act on their own, or they may act as proxies for nation-states that wish to duck responsibility. Either way, stateless groups are forces to be reckoned with.

At the same time, a different set of technology trends means that small numbers of people can obtain incredibly lethal power. Now, for the first time in human history, a small group can be as lethal as the largest superpower. Such a group could execute an attack that could kill millions of people. It is technically feasible for such a group to kill billions of people, to end modern civilization—perhaps even to drive the human race to extinction.

Our defense establishment was shaped over decades to address what was, for a long time, the only strategic threat our nation faced: Soviet or Chinese missiles. More recently, it has started retooling to address tactical terror attacks like those launched on the morning of 9/11, but the reform process is incomplete and inconsistent. A real defense will require rebuilding our military and intelligence capabilities from the ground up. Yet, so far, strategic terrorism has received relatively little attention in defense agencies, and the efforts that have been launched to combat this existential threat seem fragmented.

History suggests what will happen. The only thing that shakes America out of complacency is a direct threat from a determined adversary that confronts us with our shortcomings by repeatedly attacking us or hectoring us for decades.

Our present foes are not doing that. Instead, they are likely to wait patiently between attacks. For now, they are satisfied with tactical terrorism, but at some point, they will have the means, opportunity, and motive to turn to strategic-terror weapons.

Therefore, we will most likely continue to lumber along on our current path, addressing some issues and ignoring others. Then the terrorists will launch the next attack. With luck, we will detect it in time to prevent a major disaster, but a more likely scenario is that a strategic-terror attack in the next decade or so will kill between 100,000 and one million Americans. Then we will surely get serious about strategic terrorism.

Or we could start now.

References

1 Inglesby, T. V. *et al.* "Anthrax as a Biological Weapon, 2002: Updated Recommendations for Management," *Journal of the American Medical Association,* 287:17, 2236–2252; May 1, 2002.

2 Keim, P. *et al.* "Molecular Investigation of the Aum Shinrikyo Anthrax Release in Kameido, Japan," *Journal of Clinical Microbiology,* 39:12, 4566–4567; December 2001.

3 Wein, L. M., Craft, D. L., and Kaplan, E. H. "Emergency Response to an Anthrax Attack," *Proceedings of the National Academy of Sciences USA,* 100:7, 4346–4351; March 21, 2003.

4 Office of Technology Assessment, U.S. Congress. *Proliferation of Weapons of Mass Destruction: Assessing the Risks.* U.S. Government Printing Office, August 1999.

5 Wright, J. G. *et al.* "Use of Anthrax Vaccine in the United States," *Morbidity and Mortality Weekly Report,* 59:RR-6, 1–30; July 23, 2010.

6 Pomerantsev, A. P. *et al.* "Expression of Cereolysin AB Genes in Bacillus anthracis Vaccine Strain Ensures Protection Against Experimental Hemolytic Anthrax Infection," *Vaccine,* 15:17–18, 1846–1850; December 6, 1997.

7 Lane, J. M., and Poland, G. A. "Why Not Destroy the Remaining Smallpox Virus Stocks?" *Vaccine,* 29:16, 2823–2824; April 5, 2011.

8 Bozzette, S. A. *et al.* "A Model for a Smallpox-Vaccination Policy," *New England Journal of Medicine* 348, 416–425; January 30, 2003.

9 Kaplan, E. H., Craft, D. L., and Wein, L. M. "Emergency Response to a Smallpox Attack: The Case for Mass Vaccination," *Proceedings of the National Academy of Sciences USA*, 99: 16, 10935–10940; August 6, 2002.

10 Ramshaw, I. *XIII International Poxvirus and Iridovirus Symposium, Montpellier, France;* September 2000

11 Nowak, R. "Disaster in the making." *New Scientist;* January 13, 2001.

12 Buller, R.M.L. *Smallpox BioSecurity: Preventing the Unthinkable, Geneva, Switzerland;* October 21–22, 2003.

13 MacKenzie, D. "U.S. Develops Lethal New Viruses," *New Scientist;* October 29, 2003.

14 Imai, M. *et al.* "Experimental Adaptation of an Influenza H5 HA Confers Respiratory Droplet Transmission to a Reassortant H5 HA/H1N1 Virus in Ferrets," *Nature* 486, 420–428; June 21, 2012.

15 UNAIDS. *Global Report: UNAIDS Report on the Global AIDS Epidemic 2012.* UNAIDS, 2012.

16 Judson, O. "A Bug's Death," *The New York Times;* September 25, 2003.

Made in United States
North Haven, CT
04 November 2026

21765614R00039